TEIXOBACTIN

Teixobactin: Killing Pathogen without Detectable Resistance

INDEX

Teixobactin: Killing Pathogen without Detectable Resistance

Introduction

In 1928, **Alexander Fleming** discovered that penicillin has the ability to kill many disease causing bacteria, creating a new kind of medication known as antibiotics. Over the years since then, many variations of penicillin have been discovered and other groups or "classes" of antibiotics have also been found. Each class of antibiotics continues to be refined and new variations of them are found. But since 1987, when the group called "lipopeptides" was found, no new classes of antibiotics have been discovered – until 2015.

The lack of new antibiotics has become problematic because in the meantime, bacteria have been evolving to become resistant to many existing antibiotics. They seem to have an ability to use gene sharing to acquire resistance from other bacteria.

In January of 2015, a new antibiotic called **Teixobactin** was discovered that can kill MRSA, tuberculosis, strep, anthrax, and C-diff, all of which have developed levels of antibiotic resistance. All of these pathogens are considered to be "gram-positive" and Teixobactin is not effective against gram-negative bacteria, which *includes E. coli, cholera, legionella, gonorrhoeae, meningitis, salmonella, yersinia* and more.

Teixobactin is a newly discovered natural occurring antibiotic peptide that has potential to avoid the development of resistance. Teixobactin was discovered using a new method of culturing bacteria, called an isolation chip (ichip). The iChip technology was used to screen 10,000 bacterial isolates for their antimicrobial activity against *Staphyloccocus aureus*. The research group succeeded in isolating an active compound with mass 1,242 Da. The use of NMR together with advanced Marfey's analysis allowed to assign the sterochemical conformation of amino acid residues for the peptide termed teixobactin. The peptide is a depsipeptide that contains enduracididine, methylphenuylalanine, and four D-amino acids.

The researchers also identified the gene cluster via a homology search using sequence data from genomic DNA from *E. terrae*. Furthermore, the research group was able to show that teixobactin expressed excellent activity against Gram-positive pathogen, including drug-resistant strains.

Teixobactin: Killing Pathogen without Detectable Resistance

Identification

The iChip, a collection of plates with hundreds of wells covered by two layers of semi-permeable membranes, is used to isolate previously unculturable bacteria by growing microbes placed in the device *in situ* where their typical environmental factors are used to cultivate the bacteria.

In this case, a soil sample was diluted so that on average one bacteria cell could be introduced into each iChip well, and the device was covered by the semi-permeable membranes to be placed back into the soil where the bacteria was originally collected. Compared to standard Petri dishes, the colony count for soil bacteria using the iChip was 5 times higher when cultivating from a single colony's worth of bacteria. Extracts from thousands of isolates were tested for antibiotic activity, and the discovery of Teixobactin's properties lead to the identification and classification of **Eleftheria terrae** using 16S rRNA genesequencing. Relatives of *E. terrae* were not previously known to produce antibiotics. Though the use of the iChip's cultivating abilities, other antibiotics like teixobactin could be discovered.

Many species of bacteria won't grow well in laboratory environments. The ichip allows researchers to seal individual bacteria samples in compartments that are isolated from each other but with access to the nutrient environment found in natural soil through a permeable membrane. The collection of samples is then buried and allowed to grow naturally.

Figure: Chemical Structure

4

Teixobactin: Killing Pathogen without Detectable Resistance

This powerful new antibiotic is reported to kill an array of drug-resistant bacteria in experimental infections in mice. It is a first new class of antibiotics discovered in decades. Many scientists think that this is a much-needed breakthrough that could lead to a whole new class of disease-fighting treatments which has excited scientists, doctors and the public all over the world. Teixobactin is an extract of β-proteobacteria named *Eleftheria terrae*.

This antibiotic is only tested in Mice, to become a drug to treat infections in people, clinical trials will need to be carried out to make sure that the drug is safe and works in patients. Human trails are not yet started so drug will not be available in market at least for 3-4 years.

Time will prove Teixobactin will develop into a new drug or not. A new tool called iChip, was used for the isolation of Teixobactin from a soil microorganism *Eleftheria terrae* which does not grow in test tube in laboratory condition. This tool or device allow the researchers to dilute the bacteria containing soil samples, sandwich them between two semi-permeable membranes, and then immerse them in soil, allowing the bacteria to be grown in the lab in natural condition.

Scientists believe that this screening could be a 'game changer' for discovering new antibiotics from environmental sources, as it allows compounds to be isolated from micro-organisms in the soil that do not grow under normal laboratory conditions.

Laboratory tests have shown that Teixobactin can kill some bacteria as quickly as established antibiotics and can cure laboratory mice suffering from bacterial infections with no toxic side-effects. Studies have also revealed the prototype drug works against harmful bacteria in a unique way that is highly unlikely to lead to drug-resistance – one of the biggest stumbling blocks in developing new antibiotics.

Teixobactin: Killing Pathogen without Detectable Resistance

Uses

Teixobactin is not a panacea against all bacteria. According to the researcher, Teixobactin was ineffective against most Gram-negative bacteria but had showed excellent activity against Gram-positive pathogens, including drug-resistant strains:

1. Potency against most species, including difficult-to-treat enterococci and tuberculosis
2. Teixobactin was exceptionally active against *Clostridium difficile* and *Bacillus anthracis*
3. Teixobactin had excellent bactericidal activity against *aureus*, MRSA (Methicillin Resistant *Staphylococcus aurues* and Vancomycin Intermediate *Staphylococcus aureus* (VISA)

Teixobactin: Killing Pathogen without Detectable Resistance

Mechanism of Action

Teixobactin is an unusual depsipeptide which contains enduracididine, methylphenylalanine, and four D-amino acids. It inhibits cell wall synthesis by binding to a highly conserved motif of lipid II (precursor of peptidoglycan) and lipid III (precursor of cell wall teichoic acid).

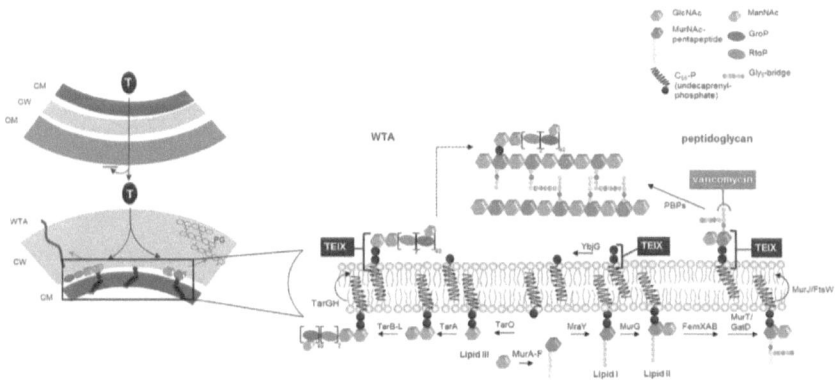

Figure: Model for the mechanism of action of Teixobactin

The antibiotic works through inhibiting cell wall synthesis by binding to two cell wall precursors, Lipid II and Lipid III. Lipid II is precursor for peptidoglycan, a polymer that forms the cell wall, and Lipid III is a precursor for techoic acid, polymers that provide rigidity to the cell wall in gram positive bacteria.

The specificity of Teixobactin to the lipid cell wall precursors and the presence of an outer cell membrane is gram negative cells explain the antibiotic's lack of effectiveness on gram negative cells. Teixobactin blocks precursor recycling as well as techoic acid and peptidoglycan synthesis.

Though researchers looked for possible Teixobactin-resistant mutants using *S. aureau* and *M. tuberculosis* through plating the bacteria in media with four times their respective MICs, no resistant mutants were found.

Teixobactin: Killing Pathogen without Detectable Resistance

Studies done in the past on Vancomycin, another antibiotic that binds to lipid cell wall precursors, suggest that Teixobactin lack of development of resistance may be due to the fact that the lipids are made from organic precursors rather than synthesized de novo from DNA like proteins, making development of resistance through mutation alone more difficult. In addition, the binding target of the drug is highly conserved among bacteria.

Resistance to Vancomycin was identified almost 40 years after the drug's discovery when it is believed that the self-resistance vector from Vancomycin -producing bacteria was captured by pathogenic bacteria through horizontal gene transfer.

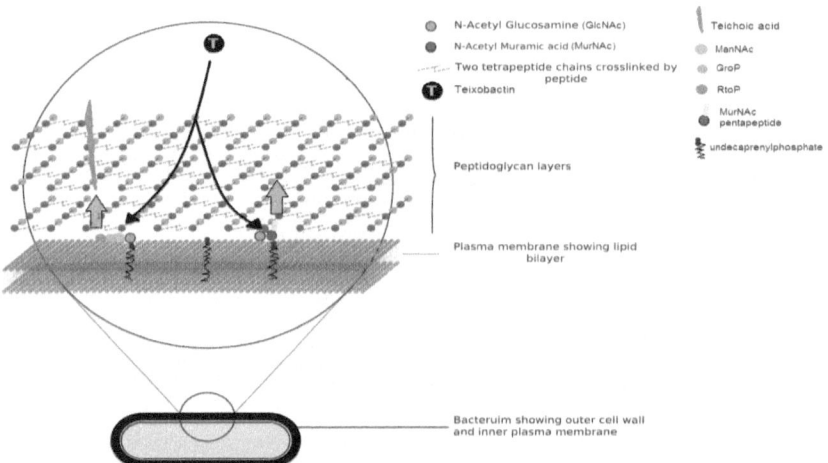

Simplified schematic of the inhibition of cell wall synthesis by Teixobactin in a gram positive bacterium

Figure: Inhibition of cell wall synthesis by Teixobactin

Teixobactin: Killing Pathogen without Detectable Resistance

How iChip Works

In a single scoop of soil, bacteria and fungi number in the millions. They also come in thousands of varieties, and survive by fighting each other. We know this because for the past century, several newly discovered antibiotics have been found by isolating them from the bacteria and fungi that produce them to defend their own lives.

The trouble, however, is that only about 1% of the microbes in the soil (or sea water) can be reliably grown under lab conditions. This means that so far we have not been able to study the remaining 99%, which are bound to produce antibiotics unknown to us.

This is the problem Kim Lewis and Slava Epstein at Northeastern University in Boston and colleagues have been busy trying to solve. After more than a decade of work, they have a solution in the iChip technology.

To make it work, a sample of soil is diluted and then poured on the iChip, which consists of hundreds of small holes. Because of the dilution, it is hoped that only one microbe is caught in each hole.

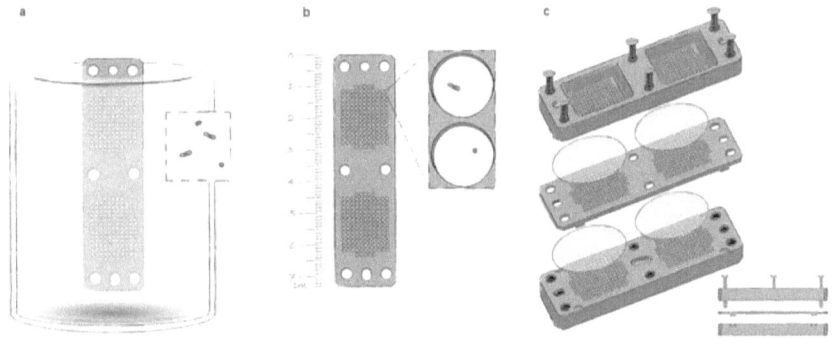

Figure: The iChip

The iChip is then covered with membranes on both sides and put back into the soil sample. The membranes contain pores that are only large enough for the chemical nutrients to flow in but small enough to block the movement of any bacteria.

This means the single bacteria in each of the holes in the iChip can consume all the nutrients it would naturally find in the environment and multiply, but not be contaminated with other bacteria in the soil.

Remarkably, it has been shown that this method can help nearly one in two bacteria to start growing in the iChip cells. Better still, three-quarters of the iChip bacteria can then be transferred to and grown in lab solutions. The improvement from the 1% that could be previously grown in labs.

Why this transition through the iChip allows previously incapable bacteria to grow in lab solutions is not clear, but it may have to do with mutations gained by the bacteria during the process. And this step is crucial because it might help to overcome a huge barrier that was stopping the development of new antibiotics – growing bacteria under lab conditions to study and isolate the antibiotics they produce.

Teixobactin: Killing Pathogen without Detectable Resistance

Discovery of Teixobactin

It has been recognised that there are only a relatively few species of microorganisms in common use in laboratories which may be sources of antibiotics. A novel culture technique was used to extend the opportunities to isolate microbes ("uncultured bacteria"?) from their natural environment - still mostly soil samples. This was achieved after burying sample cells in a variety of places for some time - several weeks.

The sample cells - iChip (isolation chips) consisted of 3 layers of plates with about three hundred matched perforations, screwed together to enclose 2 layers of a "semipermeable" film. Before burial, the central section was dipped in a liquid sample soil suspension (of a suitable dilution), likely to contain bacteria from the particular environment.

Each perforation forms a small diffusion chamber sandwiched between the membranes. These chambers allow for a free exchange of chemicals with the external environment by diffusion, while restricting the movement of cells and contamination/interaction with other microbes.

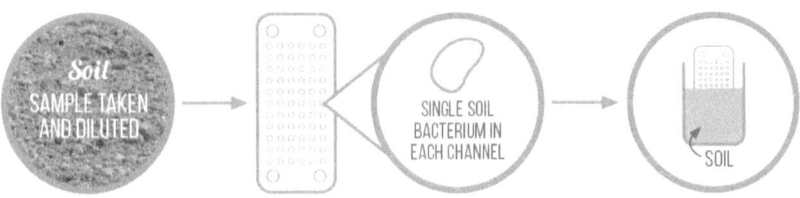

Figure: Use of iChip device to extract Teixobactin compound from soil

The small size should result in each perforation potentially containing (different?) microbial colonies, each derived from individual cells in the initial inoculum permitting the isolation of many isolates in pure culture.

It has been suggested that continuous cultivation in diffusion chambers adapts some microorganisms for growth under otherwise prohibitive conditions in the laboratory. These colonies could then be transferred to normal laboratory media (agar in Petri dishes) in the ordinary way. A large number of microbial species were obtained, which could then be cultured and further examined in the normal way in the laboratory, but specific growth factors were found to be important.

Some bacteria were found to depend on other bacterial species: they require siderophores produced by other species in order to solubilise and access iron in the surrounding medium. From several thousand isolates, 25 antibiotic-producing strains were identified as potentially useful.

The most promising one of these has been named teixobactin, derived from a newly described bacterium (within the class β-proteobacteria) provisionally named Eleftheria terrae.

Teixobactin consists of a chain of 11 amino acids, the last 4 of which are looped back forming a ring structure. At each end of the molecule are unusual substituted amino-acids: N-methyl-D-phenylalanine and enduracididine (not the usual amino acids in proteins). It is a depsipeptide - a molecule that has both peptide and ester linkages in proximity in the same amino acid chain. Minimum inhibitory concentration (MIC) was used to measure the effectiveness of teixobactin against pathogenic bacteria.

While teixobactin was not effective against gram-negative bacteria, less than 0.6 μg/mL of the drug was necessary to visibly inhibit the growth of many Gram positive bacteria tested including pathogenic MRSA, VRE, *Bacillus anthracis*, and *Clostridium difficile*. No toxicity toward mammalian cells was found as expected from the mechanism of action.

Synthesis scheme for core Teixobactin fragment

The approach used in this work involves synthesising the peptide completely on solid phase and only a single purification was required after the final cleavage. The synthesis of building block AllocHN-D-Thr-OH was also reported for first time and which can be directly coupled to the resin without protecting the amino acid side chain. 2- Chlorotritylchloride resin was chosen instead of Wang resin as the peptide could be cleaved off from the resin without removing the side chain protecting groups of the attached amino acids

(a) The first amino acid loaded on the resin in this case is Fmoc-Ala-OH followed by (b) an amide bond coupling with Alloc-D-Thr-OH. (c) Fmoc-Ile-OH is then coupled at this stage via an ester bond to the free –OH side chain of threonine. Next, (d) arginine was coupled via an amide bond, the Fmoc protecting group is removed and (e) the N-terminus is protected via a trityl protecting group (combining cleavage and deprotection in a single step) to facilitate the cleavage and cyclisation as described in reactions (h and i). (f) The Alloc group protecting the N-terminus of the threonine is then removed and (g) the peptide chain is built via standard solid phase peptide synthesis (SPPS). Partial cleavage was performed using 2 : 5 : 93 TFA:TIS :DCM followed by cyclisation using 1-[bis(dimethylamino) methylene]-1H-1,2,3-triazolo[4,5-b]-pyridinium 3-oxid hexafluorophosphate (HATU) as a coupling reagent and

Teixobactin: Killing Pathogen without Detectable Resistance

DIPEA as a base in DMF for 1 h. The protecting groups are then cleaved off using 95 : 2.5 : 2.5 TFA : TIS :H_2O yielding the desired peptide (22%recovery, refer ESI,† S10). After successful synthesis of analogue, the general applicability of this approach was tested for the synthesis of analogue 3. In analogue 3, the three D-amino acids residues (Phe, Gln and Ile) were replaced by L-amino acid residues. The synthesis of analogue 3 also worked efficiently (17% recovery).

The detailed characterisation of 1 and 3 were performed using LC-MS and NMR. The NMR spectra of product 1 was shown to be identical as reported previously.[5,9] The NOEs of 1 were characteristic of a random coil, however, the NOEs of 3 suggested a considerable degree of structure.

The analogues 1 and 3 were evaluated for their antibacterial activity. MIC results showed a trend similar to Teixobactin for analogue 1 against both Gram-positive and Gram-negative bacteria. Analogue 3 was not active against Gram-negative bacteria. Moreover, analogue 1 was 64 times more effective than analogue 3 against Gram-positive bacteria.

This difference in antibacterial activity has established that the three D-amino acids residues of 1 are critical for the antibacterial activity.

Teixobactin: Killing Pathogen without Detectable Resistance

Methods

Isolation and cultivation of producing strains- A sample of 1 g of soil sample collected from a grassy field in Maine was agitated vigorously in 10 ml of deionized H_2O for 10 min. After letting the soil particulates settle for 10 min, the supernatant was diluted in molten SMS media (0.125 g casein, 0.1 g potato starch, 1 g casamino acids, 20 g bacto-agar in 1 litre of water) to achieve an average concentration of one cell per 20 ml of medium. Then 20 ml aliquots were then dispensed into the wells of an iChip. The iChip was placed in direct contact with the soil. After one month of incubation, the iChips were disassembled and individual colonies were streaked onto SMS agar to test for the ability to propagate outside the iChip and for colony purification.

Extract preparation and screening for activity- Isolates that grew well outside the iChip were cultured in seed broth (15 g glucose, 10 g malt extract, 10 g soluble starch, 2.5 g yeast extract, 5 g casamino acids, and 0.2 g $CaCl_2N_2H_2O$ per 1 litre of deionized H_2O, pH 7.0) to increase biomass, followed by 1:20 dilution into 4 different fermentation broths. After 11 days of agitation at 29 ˚C, the fermentations were dried and resuspended in an equal volume of 100% DMSO. Then 5 ml of extracts were spotted onto a lawn of growing *S.aureus* NCTC8325-4 cells in Mueller-Hinton agar (MHA) plates. After 20 hrs of incubation at 37 ˚C, visible clearing zones indicated antibacterial activity. The extract from this isolate, which was provisionally named *Eleftheria terrae* sp., produced a large clearing zone. Although *E. terrae* sp. produced antibacterial activity under several growth conditions, the best activity (that is, largest clearing zone) was seen with R4 fermentation broth (10 g glucose, 1 g yeast extract, 0.1 g casamino acids, 3 g proline, 10 g $MgCl_2$-$6H_2O$, 4 g $CaCl_2$-$2H_2O$, 0.2 g K_2SO_4, 5.6 g TES free acid (2-[[1,3-dihydroxy-2-(hydroxymethyl)propan-2-yl]amino]ethanesulfonic acid) per 1 litre of deionized H_2O, pH 7).

Sequencing of the strain- Genomic DNA of *E. terrae* was isolated. Sequencing was performed at the Tufts University Core Facility (Boston, MA). A paired-end library with an insert size of approximately 800 bases was generated and sequenced using Illumina technology. The read length was 251 bases per read.

Strain identification- A suspension of cells was disrupted by vigorous agitation with glass beads (106nm or smaller) and the supernatant used as template to amplify the 16S rRNA gene, using GoTaq Green Master Mix (Promega M7122), and the universal primers E8F and U1510R33. The thermocycler parameters included 30 cycles of 95 °C for 30 s, 45 °C for 30 s and 72 °C for 105 s. The amplified DNA fragment was sequenced and the sequence compared by BLAST to cultured isolates in the Ribosomal Database Project. The assembled genome for *E. terrae* was submitted to the RAST genome annotation server at which produced a list of closest relatives with published genomes. These are *Alicycliphilus denitrificans, Leptothrix cholodnii, Methylibium petroleiphilum*, and *Rubrivivax gelatinosus*, and their DNA hybridization (DDH) values of these genomes to *E. terrae* were then predicted by the Genome-to-Genome Distance calculator 2.0. Note that *M. Petroleiphilum* and *R. gelatinosus* are present on the phylogeny tree of *E. terrae*

Biosynthetic gene cluster identification- By screening the draft genome of *E. terrae*, obtained by Illumina sequencing, many gene fragments putatively belonging to NRPS coding genes were identified. The assembly was manually edited and gap closure PCRs were performed. Sanger sequencing of the resulting fragments allowed the closure of the gene locus corresponding to the Teixobactin biosynthetic gene cluster. The specificity of the adenylation domains was determined using the online tool NRPSpredictor2.

Strain fermentation and purification of Teixobactin- Homogenized colonies were first grown with agitation in seed broth. After 4 days at 28 °C, the culture was diluted 5% (v/v) into the R4 fermentation media, and production monitored with analytical HPLC. For scale-up isolation and purification of teixobactin, 40 litres of cells were grown in a Sartorius Biostat Cultibag STR 50/200 Bioreactor for about 7 days. The culture was centrifuged and the pellet extracted with 10 litres of 50% aqueous acetonitrile and the suspension again centrifuged for 30 min. The acetonitrile was removed from the supernatant by rotary evaporation under reduced pressure until only water remained. The mixture was then extracted twice with 5 litres of n-BuOH. The organic layer was transferred to a round bottom flask and the n-BuOH removed by rotary evaporation under reduced pressure. The resulting yellow solid was dissolved in DMSO and subjected to preparatory HPLC (SP: C_{18}, MP:H_2O/MeCN/0.1% TFA). The fractions containing teixobactin were then pooled and the acetonitrile removed by rotary evaporation under reduced pressure. The remaining aqueous mixture was then lyophilized to leave a white powder (trifluoroacetate salt). Teixobactin was then converted to a hydrochloride salt, and endotoxin removed as follows. 100 mg of teixobactin (TFA salt) was dissolved in 100 ml of H_2O and 5 g of Dowex (1×4 Cl⁻ form) was added and the mixture incubated for 20 min with occasional shaking. A 10 g Dowex (1×4 Cl⁻ form) column was prepared and the mixture was then poured onto the prepared column and the solution was allowed to elute slowly. This solution was then poured over a fresh 10 g Dowex (1x4 Cl⁻ form) column and the resulting solution filtered through a Pall 3K Molecular Weight Centrifugal filter. The clear solution was then lyophilized to leave a white powder.

Minimum Inhibitory Concentration (MIC)- MIC was determined by broth micro diution method according to CLSI guidelines. The test medium for most species was cation adjusted Muller-Hinton broth (MHB). The same supplement with 3% lysed horse blood (Cleveland Scientific, Bath, OH) for growing Streptococci. *Haemophilus* test medium was used for *H.influenza* (Teknova, Hollister, CA) Middlebrook 7H9 broth (Difco) was used for mycobacteria, Schaedler anaerobe broth (Oxoid) was used for *C. difficle*, and fetal bovine serum (ATCC) was added to MHB (1:10) to test the effect of serum. All test media were supplemented with 0.002% polysorbate 80 to prevent drug binding to plastic surfaces and cell concentration was adjusted to approximately 5×10^5 cells per ml. After 20 hr of incubation at 37 ˚C (2 days for *M. smegmatis*, and 7 days for *M. tuberculosis*), the MIC was defined as the lowest concentration of antibiotic with no visible growth. Expanded panel antibacterial spectrum of Teixobactin was tested at Micromyx, Kalalmazoo, MI, in broth assays. Experiments were performed with biological replicates. Minimum bactericidal concentration (MCB). *S.aureus* NCTC8325-4 cells from the wells from an MIC microbroth plate that had been incubated for 20 hr at 37 ˚C were pelleted. An aliquot of the initial inoculums for the MIC plate onto MHA, and the colonies enumerated after incubating for 24 hr at 37 ˚C. The MBC is defined as the first drug dilution which resulted 99.9% decrease from the initial bacterial titre of the starting inoculums, and was determined to be 2 × MIC for Teixobactin. Experiments were performed with biological replicates.

Time-dependent killing- An overnight culture of cells (*S. aureus* HG003; Vancomycin intermediate *S. aureus* SA1287) was diluted 1:10,000 in MHB and incubated at 37 ˚C with aeration at 225 r.p.m. for 2 h (early exponential) or 5 h (late exponential). Bacteria were then challenged with antibiotics at 10 × MIC (a desirable concentration at the site of infection), oxacillin (1.5 µgml⁻¹), Vancomycin (10 µgml⁻¹) or Teixobactin (3 µgml⁻¹) in culture tubes at 37 ˚C and 225 r.p.m. At intervals, 100 µl aliquots were removed, centrifuged at 10,000g for 1 min and resuspended in 100 µl of sterile phosphate buffered saline (PBS). Tenfold serially diluted suspension were platted on MHA plates and incubated at 37 °C overnight. Colonies were counted and C.F.U per ml was calculated. For analysis of lysis, 12.5 ml of culture at A_{600nm} (OD_{600}) of 1.0 was treated with 10 × MIC of antibiotic for 24 h, after which, 2 ml of each culture was added to glass test tubes and photographed. Experiments were performed with biological replicates.

Resistance studies- For single step resistance S. aureus NCTC8325 at 10^{10} CFU were platted onto MHA containing 2×, 4×, and 10× MIC of Teixobactin. After 48 hr of incubation at 37°C, no resistant colonies were detected, giving the calculated frequency of resistance to Teixobactin of $<10^{-10}$. For M. tuberculosis, cells were cultured in 7H9 medium and plated at 10^9 cells per ml on 10 plates and incubated for 3 weeks at 37°C for colony counts. No colonies were detected.

For resistance developments by sequential passaging, *S. aureus* ATCC 29213 cells at exponential phase were diluted to an A_{600nm} (OD_{600}) of 0.01 ml of MHB supplemented with 0.02% polysorbate 80 containing Teixobactin or ofloxacin. Cells were incubated at 37°C with agitation, and passaged at 24 h intervalsin the presence of Teixobactin or Ofloxacin at subinhibitory concentration. The MIC was determined by broth microdilution. Experiments were performed with biological replicates.

Mammalian cytotoxicity- The CellTiter 96 A Queous one solution Cell Proliferation Assay was used to determine the cytotoxicity of Teixobactin. Exponentially growing NIH/3T3 mouse embryonic fibroblastic (ATCC CRL-1658, in Dulbecco's Modified Eagle's medium supplemented with 10% fetal calf serum) were seeded into a 96 well flat bottom plates, and incubated at 37°C. After 24 h, the medium was replaced with fresh medium containing test compounds (0.5 µl of twofold serial dilution in DMSO to 99.5 µl of media). After 48 h if incubation at 37°C, reporter solution was added to the cells and after 2 h, the A_{490nm} (OD_{490}) was measured using a spectra max Plus Spectrophotometer. Experiments were performed with biological replicates.

Haemolytic activity- Fresh human red blood cells were washed with PBS until the upper phase was clear after centrifugation. The pellet was resuspended to an A_{600nm} (OD_{600}) of 24 in PBS, and added to the wells of a 96 well U-bottom plate. Teixobactin was serially diluted two fold in water and added to the wells resulting in a final concentration ranging from 0.003 to 200 μg ml⁻¹. After one hour at 37 °C, cells were centrifuged at 1,000 g. The supernatant was diluted and $A_{450nm}(OD_{450})$ measured using a Spectramax Plus Spectrophotometer.

Macromolecular synthesis- S. aureus NCTC8325-4 cells were cultured in minimal medium (0.02 M HEPES, 0.002 M MgSO$_4$, 0.0001 M CaCl$_2$, 0.4% succinic acid 0.043 M NaCl$_2$, 0.5% (NH$_4$)$_2$SO$_4$) supplemented with 5% tryoticsoy broth (TSB). Cells were pelleted and resuspended into fresh minimal medium supplemented with 5% TSB containing test compounds and radioactive precursors were glucosemine hydrochloride, D-[6-³H(N)] (1mCiml⁻¹), leucine, L-[3,4,5-³H(N)] (1mCiml⁻¹), uridine [5-³H] (1mCiml⁻¹) or thymidine, [methyl-³H] (0.25mCiml⁻¹) to measure cell wall, protein, RNA and DNA synthesis respectively. After 20 min of incubation at 37°C, aliquots were removed, added to ice cold 25% trichloroacetic acid (TCA), and filtered using Multiscreen Filterplates (Millipore Cat.MSDVN6B50). The fiters were washed twice with ice cold 25% TCA, twice with ice cold water, dried and conted with scintillation fluid using Perkin Elmer MicroBeta Scintillation and Luminescence counter.

Intracellular accumulation of UDP-N-acetyl-muramic acid pentapeptide- Analysis of the cytoplasmic peptidoglycan nucleotide precursor pool was examined using S. aureus ATCC29213 grown in 25 ml MHB. Cells were grown to an $A_{600nm}(OD_{600})$ of 0.6 and incubated with 130 μl ml⁻¹ of chloramphenicol for 15 min. Teixobactin was added at 1.25 and 5× MIC and incubated for another for another 60 min. Vancomycin (VAN;10× MIC), known to form a complex with lipid II, was used as positive control. Cells were collected and extracted with boiling water. The cell extract was then centrifugal and the supernatant lyophilized. UDP-linked cell wall precursors were analysed by RP18-HPLC and confirmed by MALDI-ToF mass spectrometry.

Cloning, overexpression and purification of S.aureus UppS and YbjG as His$_6$- tag fusions- S. aureus N315 uppS (SA0415) were amplified using forward and reverse primers uppS_FW-5'-TCGGAGGAAAGCATATGTTTAAAAAGC-3',uppS_RV-5'-ATACTCTCGAGCTCCTCACTC-3',Sa0415_FW-5'-GCGCGGGATCCATGATAGATAA AAAATTAACATCAC-3' and SA0415_RV-5'-GCGCGCTCGAGAACGCGTTGTCGATG AT-3', respectively and cloned into a modified pET20 vector using restriction enzymes NdeI (*uppS*) or Bam HI (*ybjG*) and Xhol, to generate C-terminal His$_6$ fusion proteins. Reombinant UppS- His$_6$ enzyme was overexpressed and purifie as described for MurG. For overexpression and purification of YbjG His$_6$ *E.coli* BI21 (DE3)C43 cells transformed with the appropriate recombinant plasmid were grown in 2YT medium (50 µg/ml ampicillin) at 25˚C. At an A$_{600nm}$(OD$_{600}$) of 0.6 IPTG was added at a concentration of 1mM to induce expression of the recombinant of the recombinant proteins. After 16 h cells were harvested and resuspended in buffer A (25 mM Tris/HCl,pH 7.5, 150mM NaCl, 2 mM β-mercaptoethanol, 30% glycerol and 1 mM MgCl$_2$).2mg/ml lysozyme, 75 µg/ml. RNase were added cells were incubated for 1 h on ice, sonicated and the resulting suspension was centrifuge (20,000g, 30 min, 4˚C). Pelleted baxterial membranes were washed three times to remove remaining cytoplasmic content. Membrane proteins were solubilised in two successive steps with buffer. Acontaining 17.6 mM n-dodecyl-β-D-maltoside (DDM). Solubilized proteins were separated from cell debris by centrifugation (20,000g, 30 min, 4˚C) and the supernatant containing recombinant proteins was mixed with Talon-agarose (Clontech) and purification was performed. Purity was controlled by SDS-PAGE and protein concentration was determined using Bradford protein assay (Biorad).

***In vitro* peptidoglycan synthesis reactions**- *In vitro* peptidoglycan biosynthesis reaction were performed as described using purified enzymes and substrates. The MurG activity assay was performed in a final volume of 30 µl containing 2.5nmol purified lipid I, 25nmol UDP-N-acetyl glucosamine (UDP-GlcNAc) in 200mM Tris-HCl, 5.7 mM MgCl$_2$,pH 7.5, and 0.8% Triton X-100 in the presence of 0.45 µg of purified, recombinant MurG-His$_6$ enzyme. Reaction mixtures were incubated for 60 min at 30˚C. For quantitative analysis 0.5 nmol of [14C]-UDP-GlcNAc (9.25 GBq m/mol;ARC) was added to the reaction mixture. The assay for synthesis of lipid II- Gly catalysed by FemX was performed as described previously without any modification. Enzymatic activity *S. aureus* PBP2 was determined by incubating 2 nmol [14C] lipid II in 100 mM MES, 10mM MgCl$_2$,pH 5.5 in a total volume of 50 µl. The reaction was initiated by the addition of 5 µg PBP2-His$_6$ and incubated for 2.5 h at 30˚C. Monophosphorylation of C$_{55}$. PP was carried out using purified *S. aureus* YbjG-His$_6$ enzyme as describedpreviously for *E. coli* pyrophosphotase, with 0.6 µg YbjG- His$_6$ in 20 mM Tris/HCl, pH7.5, 60 mM NaCl, 0.8% Triton X-100 for 10 min at 30˚C.

In all *in vitro* assays Teixobactin was added in molar ratios ranging from 0.25 to 8 with respect to the amount of [14C] C$_{55}$.PP, lipid I or lipid II and [14C] lipid II respectively.

Synthesized lipid intermediates were extracted from the reaction mixture with n-butanol/pyridine acetate, pH 4.2 (2:1 vol/vol) after supplementing the reaction mixture with 1 M NaCl and analysed by thin-layered chromatography (TLC). Quatification was carried out using phosphoimaging (storm imaging system, GE Healthcare) as described.

Synthesis and purification of lipid intermediates- Large scale synthesis and purification of the peptidoglycan precursor lipid I and II was performed. Radio labelled lipid II was synthesized using [^{14}C]-UDP-GlcNAc (9.25 GBq m/mol;ARC) as substrate. For synthesis f the lipid II variant with a terminal D-Lacresidue,UDP-MurNAc depsipeptide (Ala-Glu-Lys-Ala-Lac) was purified from *Lactobacillus casei* ATCC393. Briefly, *L.casei* was grown in MRS broth to an A_{600nm}(OD$_{600}$) of 0.6 and incubated with 65 µg/ml of chloramphenicol for 15 min. Intracellular accumulation was achieved by incubation with Bacitracin (10 × MIC, 40 µg/ml) in the presence of 1.25 mM zinc for another 60 min. For synthesis of lipid II ending D-Ala-D-Ser the UDP-MurNAc pentapeptide (Ala-Glu-Lys-Ala-Ser) was used. The wall teichoic acid precursor lipid III was prepared using purified TarO enzyme. In short, purified recombinant TarO protein was incubated in the presence of 250 nmol C_{55} P, 2.5 µmol of UDP-GlcNAc in 83 mM Tris-HCl (pH 8.0), 6.7 mM MgCl$_2$, 8.3% (v/v) dimethyl sulfoxide, and 10 mM N-lauroyl sarcosine. The reaction was initiated by the addition of 150 µg of TarO-His$_6$ and incubated for 3 h at 30˚C. Lipid intermediate were extracted from the reaction mixture with *n*-butanol/ pyridine acetate (pH4.2)(2:1; vol/vol), analysed by TLC and purified. C_{55} P and C_{55} PP were purchased from Larodan Fine Chemicals, Sweden. [^{14}C] C_{55} PP was synthesized using purified *S. aureus* UppS enzyme based on a protocol elaborated for *E.coli* undecaprenyl pyrophosphate synthase. Synthesis was performed using 0.5 n/mol [^{14}C] famesyl pyrophosphate synthase (ARC; 2035 GBq m/mol), 5 n/mol isopentenyl pyrophosphate 0.1% Triton X-100. After 3 h of incubation at 30˚C radiolabelled C_{55} PP was extracted from the reaction mixture with BuOH and dried under vacuum. Product identity was confirmed by TLC analysis.

Antagonization assays- Antagonization of the antibiotic activity of Teixobactin by potential target molecules was performed by an MIC based setup in microtitre plates. Teixobactin (8 × MIC) was mixed with potential HPLC- purified antagonists (C_{55} P famesyl PP [C_{15} PP, Sigma Aldrich] C_{55} PP, UDP-MurNAc-pentapeptide, UDP-GlcNAc [Sigma Aldrich], lipid I, lipid II, lipid III) at a fixed molar ratio or at increasing concentration with respect to the antibiotic, and the lowest ratio leading to complete antagonization of Teixobactin ativity was determined. *S. aureus* ATCC29213(5×10^5 c.f.u per ml) were added and samples were examined for visible bacterial growth after overnight incubation. Vancomycin ($8 \times$ MIC) was used as a control.

Complete formation of Teixobactin- Binding of Teixobactin to C_{55} P, C_{55} PP, lipid I, lipid II, lipid II-D-Ala-D-Ser, lipid II-D-Ala-D-Lac and lipid III was analysed by incubating 2nmol of each purified precursor with 2-4 nmole of Teixobactin in 50 mM Tris/HCl, pH7.5, for 30 min at room temperature. Complex formation was analysed by extracting unbound precursor from the reaction mixture with n-butanol/pyridine acetate (pH4.2) (2:1 vol/vol) followed by TLC analysis using Chloroform/methanol/water/ammonia (88:48:10:1,v/v/v/v) as the solvent and detectionof lipid-containing precursors by phosphomolybdic acid staining.

hERG inhibition testing- Teixobactin was tested for inhibition of hERG activity using an IonWorksTM HT instruments, which performs electrophysiology measurements in a 384 well plate. Chinese hamster ovary (CHO) cells stably transfected with hERG were prepared as a single cell suspension in extracellular solution and aliquots added to each well by applying a vacuum beneath the plate to form an electrical seal. The resistance of each seal was measured via a common ground-electrode in the intracellular compartment and individual electrodes placed into each of the upper wells.

Cytochrome P450 inhibition- Teixobactin and control compounds were incubated with human liver microsomes at $37°C$ to determine their effect on five major human cytochrome P450 (CYP). The final microsomal concentration was 0.5 mg/ml. NADPH was added last to start the assay. After ten minutes of incubation, the amount of probe metabolite in the supernatant was determined by LC/MS/MS using an Agilent 1200 HPLC and a CTC PAL chilled autosampler, allcontrolled by MassHunter software.

***In vitro* genotoxicity**- Teixobactin was tested in an *in vitro* microsomal test that employs fluorescent cell imaging to assess cytotoxicity and quantify micronuclei. The assay was performed with CHO-KI cells in the presence or absence of Arocler treated rat liver S9 fraction to determine if any genotoxic metabolites are produced. No evidence of genotoxicity was observed with genotoxicity upto 125 µg/ml under either condition.

DNA binding- Compounds were serially diluted and mixed with sheared salmon sperm DNA (6.6 mg/ml final concentration). An aliquot was spotted onto a lawn of growing *S. aureus* NCTC8325-4 cells, and the zones of growth inhibition measured after 20 h of growth at 37˚C. A reduction in the inhibition zone size in the presence of DNA would indicate loss of antibacterial activity due to binding to the DNA.

Plasma protein binding- Protien binding of Teixobactin in rat plasma ws determined using a Rapid Equilibrium Dialysis with LC/MS analysis. Teixobactin and rat plasma in 5% dextrose containing 0.005% polysorbate 80 were added to one side of the single use RED plate dialysis chamber having an 8 kD M W cut off membrane. Following four hours of dialysis the samples from both the sides were processed and analyzed by LC/MS/MS. The Teixobactin concentration was determined, and the percentage of compound bound to protein was calculated. Teixobactin exhibited 84% plasma protein binding.

Microsomal stability- The metabolic stability of Teixobactin was measured in rat livermicrosomes using NADPH Regeneration System by monitoring the disappearance of the compound over an incubation period of two hours. Teixobactin (60 µg/ml) or Verapamil (5 µM) serving as positive control were added to 1 mg/ml microsomes at 37˚C. Aliquots were removed at 0 h, 5 h, 1 h and 2 h, and the reaction stopped by addition of 3 volumes of ice-cold acetonitrile.

Animal studies- All animal studies were carried out at Vivisource Laboratories, and University of North Texas Health Science Center and conformed to institutional animal care and use policies. Neither randomization nor blinding was deemed necessary for the animal infection models, and all animals were used. All animal studies were performed with female CD-1 mice, 6-8 weeks old.

Pharmacokinetic analysis- CD- 1 female mice were injected intravenously with a single dose of 20 mg/kg in water and showed no adverse effects. Plasma samples were taken from 3 mice per time point. An aliquot of plasma sample or calibration sample was mixed with three volumes of methanol containing internal standard incubated on ice for 5 min, and centrifuged. The protein-free supernatant was analysed by LC/MS/MS using an A gilent 6410 mass spectrometer coupled with an A gilent 1200 HPLC and a CTC PAL chilled autosampler, all controlled by MassHunter sothware. After separation on a C18 reverse phase HPLC column using an acetonitrile water gradient system, peaks were analysed by mass spectrometry using ESI ionization in MRM mode. The mean plasma concentration and the standard deviation from all 3 animals within each time point were calculated. PK parameters of test agents calculated with a non-compartmental analysis model based on WinNonlin. The mean plasma concentration from all 3 mice at each time point were used in the calculation.

Mouse sepsis protection model- Teixobactin was tested against clinical isolate *S.aureus* MRSA ATCC33591 in a mouse septicaemia protection assay to assess its *in vivo* bioavailability and PD_{50} (protective dose resulting in 50% survival of infected mice after 48h). CD-1 female mice were infected with 0.5 ml of bacterial suspension (3.28×10^7 c.f.u. per mouse) via interpretationeal injection. At one hour post-infection, mice (6 per group) were treated with Teixobactin at single intravenous doses of 20,10,5, 0.25 and 0.1 mg per kg.

Mouse thigh infection model- Teixobactin was tested against MRSA ATCC33591 in a neutropenic mouse thigh infection model. Female CD-1 mice were rendered nuetropenic by cyclophosphamide (two consecutive doses of 150 and 100 mg per kg delivered on 4 and 1 days before infection). Bacteria were resuspended in sterile saline, adjusted to an $A_{625nm}(OD_{625})$ of 0.1 mlinoculum (2.8×10^5 c.f.u. per mouse) injected into the right thighs of mice. At 2 h post-infection, mice received treatment with Teixobactin at 1,2.5,5,10 or 20 mg per kg administered in a single dose, intravenously injection (four mice per group). One group of infected mice was euthanized and thighs processed for c.f.u. to serve as the time of treatment controls. At 26 h post-infection mice were euthanized by CO_2 inhalation. The right thighs were aspectically removed , weighed, homogenized, serially diluted, and plated on trypticase soy agar for c.f.u titres.

Mouse lung infection model- Teixobactin was tested against *Streptococcus pneumonia* ATCC 6301 (UNT012-2) in an immunocompetent mouse pneumonia model to determine the compound's potential to treat acute respiratory infections. CD-1 mice were infected intranally (1.5×10^6 c.f.u. per mouse). The compound was delivered intravenously at 24 and 36 h post-infection, whereas amoxicillin was delivered at doses ranging from 0.5 to 10 mg per kg per dose (5 mice per dose). At 48 h post- infection, treated mice were eutharized, lungs aspectically removed and processed for c.f.u. counts.

Teixobactin: Killing Pathogen without Detectable Resistance

Studies carried out on Teixobactin

Teixobactin had excellent activity against Gram-positive pathogens, including drug-resistant strains Potency against most species, including difficult-to-treat enterococci and *M. tuberculosis* was below 1 µg/ml. Teixobactin was exceptionally active against *Clostridium difficile* and *Bacillus allnthracis* (minimal inhibitory concentration (MIC) of 5 and 20 ng/ml, respectively). Teixobactin had excellent bactericidal activity against *S. aureus* was superior to vancomycin in killing late exponential phase populations and retained bactericidal activity against intermediate resistance *S. aureus* (VISA) Note that frequent clinical failure inpatients with *S. aureus* MRSA treated with Vancomycin has been linked to the poor bactericidal activity of this compound. Teixobactin was ineffective against most Gram-negative bacteria, but showed good activity against a strain of *E. coli asmB1* with a defective outer membrane permeability barrier resistant to teixobactin even when plating on media with a low dose (4 × MIC) of the compound. Serial passage of *S. aureus* in the presence of sub-MIC levels of teixobactin over a period of 27 days failed to produce resistant mutants as well This usually points to a non-specific mode of action, with accompanying toxicity. However, teixobactin had no toxicity against mammalian NIH/3T3 andHepG2 cells at 100 µg/ml. The compound showed no haemolytic activity and did not bind DNA. In order to determine specificity of action of teixobactin, we examined its effect onthe rate of label incorporationinto themajor biosynthetic pathways of *S. aureus*. Teixobactin strongly inhibited synthesis of peptidoglycan, but had virtually no effect on label incorporation into DNA, RNA and protein. This suggested that teixobactin is a new peptidoglycan synthesis inhibitor. Resistance has not developed to this compound, suggesting that the target is not a protein. The essential lack of resistance development through mutations has been described for vancomycin which binds lipid II, the precursor of peptidoglycan. We reasoned that Teixobactin could be acting against the same target. Treatment of whole cells of *S. aureus* with teixobactin (1–53MIC) resulted in significant accumulation of the soluble cell wall precursor undecaprenyl-N-acetylmuramic acid-pentapeptide (UDP-MurNAc-pentapeptide), similar to the vancomycin-treated control cells, showing that one of the membrane-associated steps of peptidoglycan biosynthesis is blocked. Teixobactin inhibited peptidoglycan biosynthesis reactions in vitro in a dose-dependent manner with either lipid I, lipid II or undecaprenylpyrophosphate as a substrate. Quantitative analysis of the MurG-, FemX-, and PBP2-catalysed reactions using radiolabelled substrates, showed an almost complete inhibition at a twofold molar excess of teixobactin with respect to the lipid substrate The addition of purified lipid II prevented teixobactin from inhibiting growth of *S. aureus*. These experiments showed that teixobactin specifically interacts with the peptidoglycan precursor, rather than interfering with the activity of one of the enzymes. In order to evaluate the minimal motif required for high affinity binding of teixobactin, the direct interaction with several undecaprenyl-coupled cell envelope precursors was investigated. Purified precursors were incubated with teixobactin at different molar ratios, followed by extraction and subsequent **thin-layer chromatography analysis**.

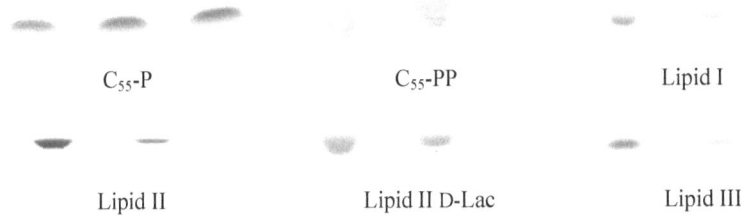

C_{55}-P	C_{55}-PP	Lipid I
Lipid II	Lipid II D-Lac	Lipid III

Fig: Complex formation of teixobactin with purified cell wall precursors. Binding of Teixobactin is indicated by a reduction of the amount of lipid intermediates

In agreement with the results obtained from the in vitro experiments, lipid I and lipid II were fully trapped in a stable complex that prevented extraction of the lipid from the reaction mixture in the presence of a two fold molar excess of the antibiotic, leading to the formation of a 2:1 stoichiometric complex. Teixobactin was active against vancomycin-resistant enterococci that have modified lipid II (lipid II-D-Ala-D-Lac or lipid II-DAla-D-Ser instead of lipid II-D-Ala-D-Ala) This suggested that, unlike vancomycin, teixobactin is able to bind to these modified forms of lipid II. Indeed, teixobactin bound to lipid II-D-Ala-D-Lac and lipid II-D-Ala-D-Ser. Moreover, teixobactin efficiently bound to the wall teichoic acid (WTA) precursor undecaprenyl-PP-GlcNAc (lipid III). Although WTA is not essential per se, inhibition of late membrane-bound WTA biosynthesis steps is lethal due to accumulation of toxic intermediates.

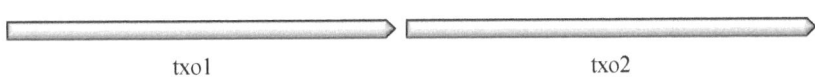

txo1	txo2

Fig: The two NRPS genes

Teixobactin: Killing Pathogen without Detectable Resistance

Table 1 | Activity of teixobactin against pathogenic microorganisms

Organism and genotype	Teixobactin MIC (μg/ml)
S. aureus (MSSA)	0.25
S. aureus 110% serum	0.25
S. aureus (MRSA)	0.25
Enterococcus faecalis (VRE)	0.5
Enterococcus faecium (VRE)	0.5
Streptococcus pneumoniae (penicillinR)	≤ 0.03
Streptococcus pyogenes	0.06
Streptococcus agalactiae	0.12
Viridans group streptococci	0.12
B. anthracis	≤ 0.06
Clostridium difficile	0.005
Propionibacterium acnes	0.08
M. tuberculosis H37Rv	0.125
Haemophilus influenza	4
Moraxella catarrhalis	2
Escherichia coli	25
Escherichia coli (asmB1)	2.5
Pseudomonas aeruginosa	>32
Klebsiella pneumonia	>32

Teixobactin: Killing Pathogen without Detectable Resistance

Microbiological Test

I.

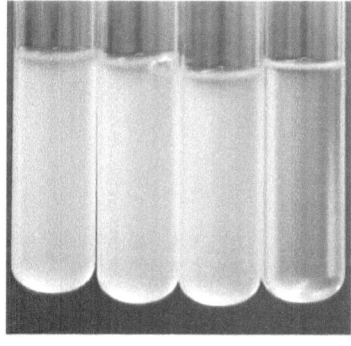

a.   b.   c.   d.

The MIC was determined by broth microdilution.

MSSA, methicillin-sensitive *S. aureus*;

VRE,vancomycin-resistant enterococci. Teixobactin

treatment resulted in lysis.

    a) = Control

    b) = Ofloxacin

    c) = Vancomycin

    d) = Teixobactin

II.

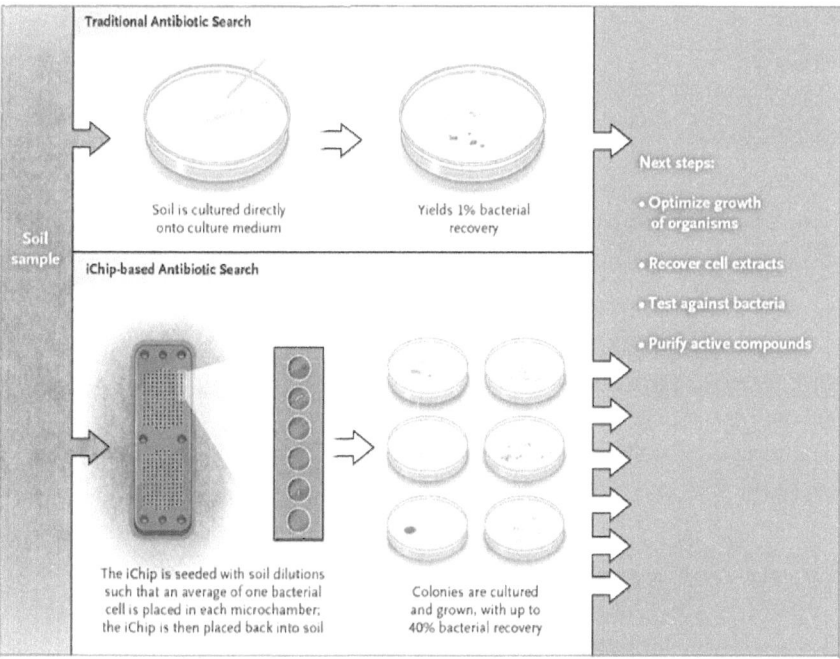

Fig: Two Methods of Culturing Microorganisms from Soil

Teixobactin: Killing Pathogen without Detectable Resistance

The traditional search for antibiotic agents involves culturing soil directly onto culture medium (e.g., an agar plate), which detects an estimated 1% of organisms present. Ling and colleagues4 used an isolation chip (iChip).5 After dilutions of soil are inoculated so that approximately one bacterial cell goes into each agar-filled chamber, the device is placed back in the soil. Many more bacteria survive and grow in the iChip than do on a traditional agar plate and, once established, are more likely to grow in vitro

In vivo Efficacy

Given the attractive mode of action of this compound, we investigated its potential as a therapeutic. The compound retained its potency in the presence of serum, was stable, and had good microsomal stability and low toxicity. The pharmacokinetic parameters determined after i.v. injection of a single 20 mg per kg dose in mice were favourable, as the level of compound in serum was maintained above the MIC for 4 h.

Ananimal efficacy study was then performed in a mouse septicemia model. Mice were infected intraperitoneally with methicillin-resistant S. aureus (MRSA) at a dose that leads to 90% of death. One hour post-infection, Teixobactin was introduced i.v. at single doses ranging from 1 to 20mg per kg. All treated animals survived, and in a subsequent experiment the PD_{50} (protective dose at which half of the animals survive) was determined to be 0.2mg per kg, which compares favourably to the 2.75 mg per kg PD_{50} of vancomycin, the main antibiotic used to treat MRSA.

Teixobactin was then tested in a thighmodel of infection with *S. aureus* and showed good efficacy as well. Teixobactin was also highly efficacious in mice infected with *Streptococcus pneumoniae*, causing a 6 log10 reduction of c.f.u. in lungs.

Clinical Trial and Resistance

Teixobactin's dual mode of action and binding to non-peptide regions suggest that resistance will be very difficult to develop. In the Laboratory, Teixobactin was effective at combating some notoriously difficult bacteria. It is only on two years on human trial but needs for confirmation over the course of five or six years.

It shows potent killing against a broad panel of bacterial pathogen including MRSA and vancomycin resistant enterococci (VRE). Uncultured organisms have recently been reported to produce interesting compound with new structures/mode of action such as Lassomycin (An inhibitor of essential mycobacterial protease) and Teixobactin (New cell wall inhibitor).

Resistance has not been developed to this compound suggesting that the target is not a protein. Resistance to vancomycin was identified almost 40 years after the drug's discovery which is believed that self-resistance vector from vancomycin producing bacteria was captured by pathogenic bacteria through horizontal gene transfer Although resistance to Teixobactin was difficult to manufacture in lab, resistance could eventually emerge in the same manner vancomycin resistance emerged, through horizontal gene transfer.

As *E. terrae* is gram negative it does not carry genes for resistance like vancomycin producing bacteria, the genes for resistance would likely comes from other soil bacteria.

Pharmacodynamic Analysis-
Teixobactin inhibits bacterial cell wall synthesis primarily acting by binding to lipid II–precursor to peptidoglycan and lipid III–precursor of cell wall teichoic acid leading to lysis of vulnerable bacteria. So there is excellent bactericidal activity. This is similar to the mechanism of action of Vancomycin.
Teixobactin forms a complex by binding to lipid I, II,and III by and incubating 2 nmol of each purified precursor with 2 to 4 n mole of Teixobactin for 30 min at room temperature. Teixobactin and control compounds like vancomycin/ lassomycin were incubated with human liver microsome at 37°C to determine their effect on five major Cytochrome p450s.

Activity against Gram negative bacteria-
According to the WHO's report in April 2014,9 one of the major global concerns of physicians is antibiotic resistance in Gram negative bacteria such as *Escherichia coli* and *Klebsiella spp.*
The Gram-negative bacterial cell envelope structure makes it difficult for many antibiotics to gain entry into the bacterium and once inside many antibiotics are exported by multidrug efflux pumps showed that teixobactin had no activity against *E. coli*, suggesting that *E. coli* is impermeable to this agent or it is effluxed (or both).
Either way, teixobactin does not inhibit *E. coli* and so is unlikely to be effective against other Gram-negative bacteria.

Will Teixobactin be developed into a new drug ?

For teixobactin (and any new compound with antimicrobial activity) to become a drug to treat infections in people, clinical trials will need to be carried out to make sure that the drug is safe, well tolerated and efficacious in patients. To do this, Teixobactin will need to be formulated so that the antibiotic remains active *in vivo* at clinically relevant sites of infection. Full toxicology tests will also need to be carried out to ensure that there are no adverse reactions or drug–drug interactions following administration of teixobactin. NovoBiotic Pharmaceuticals owns the novel chemical entities produced by the iChip and it has been stated that the hope is that teixobactin will be ready for a clinical trial in 2017. Whether it will be fully developed as a new drug remains to be seen, not least because it is questionable whether more drugs against Gram-positive bacteria are required. However, as teixobactin is active against M. tuberculosis, it could offer the opportunity for a new treatment for patients with TB. Teixobactin may also fulfil the requirements for approval by the FDA under the qualified infectious disease product (QIDP) framework, as envisaged in the USA Generating Antibiotic Incentives Now (GAIN) Act, so it could be licensed quickly. Even if teixobactin itself cannot be turned into a new drug, it is probably the first of a series of new antibiotics in its class.